Inspirational Apartments

Inspirational Apartments

Decoration Empire

TERRA

I

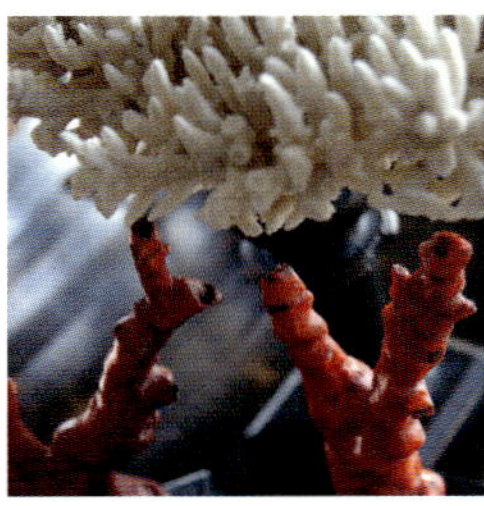

The owners: he is the retired financial director of a major Dutch insurance company; she has a passion for cooking. They swapped their large family house for this more manageable 168 m^2 apartment.

II

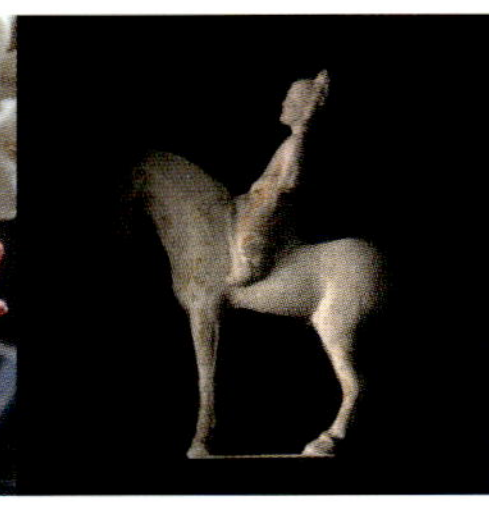

the art of art

Nothing is more personal than one's choice of art in the home - from sculpture to paintings and prints. The challenge in many cases is to mix pieces from different eras creating, not chaos, but a coherent, beautiful whole.

III

The owners: he is partner at an international law firm; she is a management coach. They wanted to create a warm, light-filled home, with a style influenced by their love of France and Italy.

IV

the art of interpretation

What a client says he or she wants from a home and what they in fact want are often two very different things - it's up to the designer to listen, ask all the right questions, and then 'read between the lines'.

V

The owners: he is the retired director of a cocoa company; they own houses in South Africa and Switzerland. They wanted to create a beautiful, restful home in which they could recharge their batteries in between trips abroad.

VI

the art of flow

A look at what flow actually means within the context of home design, why it is so important to create it as subtly and unobtrusively as possible, and what effect it has on the ambience of any home.

Contents

Introduction

Creating a truly outstanding apartment takes very special skills. It's about so much more than merely dressing walls and windows, placing furniture and artefacts, deciding on what goes where, and how. For Anne Noordam and Thong Lei at Decoration Empire, it's about creating a unique balance between certain essential elements: first, the establishing of clear communication and trust between client and designer, so that the latter can interpret what the former really wants from his or her home. From here, a layout for the apartment will be decided, with a blend of colour, scale and flow – that special something that gives a home momentum and its own special energy – determining the apartment's style. At the same time, and as a continuing part of the process, art, antiques, fabrics and wallpapers will have been selected with the utmost care, and with the utmost regard for the client's needs, passions and lifestyle. Of course, written down, the whole process sounds very calculated and possibly contrived, and it's undeniably true that there are certain rules that the best interior designers will always follow. Part of the secret of Decoration Empire's success, however, is built on the company's willingness to throw the rule book out of the window, to trust their clients' and their own instincts and to embrace the motto 'no guts, no glory'. The result, as the following pages reveal, is timeless, elegant, thought-provoking apartments; homes that – like the people who live in them and the team that designed them – are truly original, truly special.

I

Cosy apartment | 168 m². Length of project: 15 months. Contents: bedroom, bathroom, living room, kitchen-cum-dining room, storage room, integral terrace.

Small but perfectly formed

The challenge in this Voorburg apartment was to maximise the sense of light and space, and to create a kitchen as the very heart of the home, one where the lady of the owner-couple can indulge her passion for all things culinary and where the gentleman can truly unwind and relax. While the views from this second-floor home, located in a purpose-built block, are idyllic – encompassing the sweeping grounds of an 18th-century manor house that's still standing and looking towards the river – overhanging balconies surrounding each apartment swallow some of the light that might otherwise have entered the building. Meanwhile, the fact that the windows themselves are relatively small only adds to the challenge of creating the illusion of a light-filled home. Decoration Empire created the entire plan for the apartment, devised to inject light and a sense of flow into the 168-metre space. Explains Thong Lei, "Essentially what we're able to do in any situation – large project or small – is to take the hassle of re-jigging a place away from the client. We'll project-manage for them, enabling them to get on with their lives. And it's our job, too, to make sure that every centimetre of space is fully utilised, so it can achieve both its practical and its visual potential."
Simple but highly effective modifications made to this apartment include the addition of bespoke doors, hand-crafted from hardwood and fitted with nickel handles. For the ultimate

finish, every door features hinges specially made to Decoration Empire's own design. It's this sort of attention to detail that immediately elevates the look and tone of the apartment as a whole, creating a more polished and elegant effect, with a coherent visual result.

Where space is at a premium, it's important to ensure rooms are multi-functional for maximum flexibility. That's one reason it was decided to merge the kitchen with the dining room, rather than creating a separate dining space. Another key issue in the creation of any successful interior-design scheme, of course, involves an understanding of the lifestyle led by those who inhabit the space, and any good design team will first establish exactly how their clients like best to live and what they want out of their home. Thus, because the trend for separate dining rooms has died out in many homes – including this one – an intimate and inviting eating-cooking space was devised as the alternative. Here, bespoke wooden units in Havana oak and topped with honed granite create a focal point within the room. The specially-made oak dining table contrasts with the darker wood of the central island, preventing the look from becoming, in terms of colour, too predictable. Because comfort is the ultimate luxury in any home, padded dining chairs were selected, then upholstered in an informal hopsack fabric – creating the sort of atmosphere that will encourage the owners and their guests to relax and linger longer around the table.

A pale, warm and fresh colour palette in this home makes for an airy feel, with window treatments limited to simple Roman blinds that allow as much natural light in as possible. Meanwhile, artificial lighting has been specially designed to create subtle effects and to highlight focal points within each room – in the living room, for instance, a crystal lamp topped with a 'Dior'-pleated silk shade accentuates the selection of carefully-chosen brass sculptures on the table, diffusing a soft glow around them. Helping to maximise the sense of space in the apartment is a carefully controlled use of pattern – restricted to small-scale stripes and coral-effect weave on furniture upholstery – because in smaller areas it can make for a cluttered look. Instead, interest is added via texture – the touch of a faux-fur throw, the soft pile of the velvet on a small armchair – and accessories. The breadth of those used here covers everything from small classical busts to modern bronzes, to 20th-century furnishing classics, such as Louis Durot's Spiral chair in polyurethane, which dates from the 1970s. Artwork, meanwhile, includes original works by the Dutch painter Karel Appel. The eclectic nature of this selection naturally reflects the owners' tastes and passions, as well as their love of travel and appreciation of other cultures.

References to the natural world will always succeed in making a home seem airy, simply because it links the indoor space to the world beyond its four walls. In the kitchen-diner, a miniature tree provides greenery; elsewhere, it's fossilised shells, corals and anemones that pay homage to the natural world.
Every room needs a focus, and in the living room it comes courtesy of an elaborate hand-carved, gold-encrusted Chinese table that dates from the 18th century. Because of the understated nature of the rest of the room, however, the piece serves, not to overpower but, rather, to surprise and delight. In the dressing room, the wow factor is undoubtedly provided by the magnificent wall-to-wall cupboard doors in two-tone Zebrano wood, designed, hand-crafted and installed by Decoration Empire. With their pared-down finish and very simple handles, they give the room a geometric, 1960s' feel that's offset by the oak mosaic floor. It's a similarly elegant feel in the eau-de-nil-coloured bathroom, where wenge and marble surrounds are offset by a semi-circular, silk-shaded wall-light – another unexpected touch, and one that prevents the room from feeling clinical or impersonal.

This is the sort of apartment in which, as Anne Noordam puts it, "Space is a state of mind as much as anything." In other words, because the place now feels spacious and airy, yet lived in and enjoyed, it doesn't seem as if its proportions are anything less than generous. It just goes to show that, in any home, it isn't about the space at the owners' disposal, it's about how they choose to utilise that space. With the sort of imagination and skill that has been demonstrated in this project, even less-than-large homes can be made to feel capacious, welcoming and big-hearted. |

ONIONS

"Have nothing in your house that you do not know to be useful, and believe to be beautiful."

William Morris, British designer, 1834-1896

The art of **Art**

An exquisite painting or sculpture can provide the perfect inspiration or finishing touch for any room's design. Selected with care, art expresses more about mood and taste than anything else in the home, and when well-chosen pieces from different periods, are used in combination, the results can be stunning.

Of all the elements in a home, it's arguably art that is most firmly intertwined with each of the others, that is the one essential lynchpin. Art is connected to flow, because, at its most powerful, it can exude any sort of emotion. It can be soothing, stimulating, dramatic, invigorating – whatever, each piece will have an energy and momentum that's unique to itself. It's linked to colour, too, because if it doesn't harmonise at some level with other pieces within a home, it's going to stand out for all the wrong reasons. Often, too, a work of art - a painting or a piece of porcelain – can be the inspiration for the look and feel of an entire room scheme, and, as such, its importance can't be underestimated. It also benefits from a designer's ability to interpret his client's tastes correctly, because nothing is more personal than the choice of art for a home, and nothing says more about a person's taste and influences than that choice.

But selecting art is an undertaking that many feel 'unqualified' to accept. They fail to realise that the only qualification required when it comes to choosing art for the home is the courage of one's convictions. Alternatively, they may be so entrenched in their perception of what they like and what they don't like that they're not prepared to look beyond their usual tastes. With guidance from a good interior designer, however, their eyes can be opened to all sorts of different styles and genres – perhaps contemporary sculpture if they've always favoured classic paintings – that might never have appealed to them before. And the great thing about using art in the most imaginative homes is it's not an 'either/or' situation: the apartments within these pages show brilliantly how, if pieces are chosen with passion and put together with skill, a home can encompass everything from a two million BC stone statue to a 21st-century abstract canvas, with plenty of room for the neoclassical, Gustavian and Golden Age in between. These, at first glance, very different pieces can work together when they're specifically chosen because they're similar in form or in the energy that they exude, or, alternatively, they provide a beautiful contrast with one another. Again, though, Thong Lei would say that the ability to select pieces that work together comes, not through talent or study or even with time, but through 'gut instinct'.

And the only true Golden Rule when it comes to art in the home is to only buy and display what you love. Says Anne Noordam, "Art's primary purpose is to make you think, to give you pleasure. If you're not going to look at something every day that's hanging on your wall and think, 'now, *that's* glorious,' what's the point in having it?"

Finally, art is closely entwined with scale: while a vast canvas can't fail to make a dramatic statement in a room, the smallest portrait or sculpture can prove equally mesmerising, its power all the more marked in relation to its size. Equally, a picture or work of art can provide balance and scale to other elements in a room, tying them together for a coherent look. For example, a large console table may seem too dominating in a room until it is balanced by a trio of small sculptures that sit on top of it, providing an

essential focal point and anchoring the table within the room.

When it comes to wall art, something that can make or break the impact of a piece is its frame. Anne Noordam is an expert on the subject, having started her art career working on period mounts and frames. She says it's extraordinary how easy it is to dismiss a work of art, then change its frame and see it in a totally new light; "Get it wrong, and it's like seeing a beautiful woman with an unflattering haircut or ill-fitting hat; get it right, and the work of art looks as if it's inseparable from its setting." At Decoration Empire, skilled craftsmen are able to emulate the techniques of their artisan forefathers with tremendous accuracy, hand-carving and gilding authentic, appropriate frames for period works; frames that, in their original versions, might not be available on the antiques market in suitable dimensions for a given painting. But it's not just about ensuring a work has the perfect frame, it's also about knowing when to leave a piece of art exactly as it is. There are times when the only statement that's required is that made by the painting itself, and a frame would only detract from its power. This seems to be the case, in particular, with modern works of art, which often look best presented as simple canvases. Lighting, too, will have a particular effect on a painting, which is why the best interior designers will work on a scheme that enables each work to be viewed at its very best. And no attention to detail will be spared: if shades need to be specially made and lined with coloured silk to create just the right flattering glow to offset an artefact or picture, so be it.

Making art work within a home, then, is primarily dependent on a delicate balance of factors: the ability to select pieces that complement one another and their surroundings, the ability to display them in the best possible way – via shelving, hanging and lighting – but above all, for the owner of the art

to be completely true to his or her tastes. If they love what they're living with, there's a good chance that others who come into their home will respond positively to it, too.

III

Warm and sophisticated apartment | 320 m². Length of project: 14 months.

Contents: three bedrooms, two bathrooms, dressing, kitchen with informal dining area, large study for two, den and living room.

Warmth and Light

Combining French, Italian, English and Scandinavian influences, this apartment, located in a suburb of The Hague, is as warm and cosmopolitan as the couple who live in it. Decoration Empire's brief was to create a home where the owners could enjoy the ultimate standard of living, but without the hassle that running a large house entails. And with two homes to maintain abroad, they were more than happy to downsize, in Holland, from a villa to an apartment. This one – located in a mock-Tudor-style building that dates from 2005 – offered a blank canvas on which the designers, working closely with the couple, could create an elegant, welcoming home.

All that the apartment offered when work began was one 320m² space, devoid of rooms. This meant that work could begin from scratch, with the bespoke planning of the entire layout and flow of the space. Owners and designers worked closely together to decide which rooms they wanted and where, and what sort of emphasis those rooms should be given. "Flow is so vital in any home, but particularly in apartments, where space can be limited," says Design Empire's Anne Noordam. "It's what gives a place its sense of movement, its energy and momentum, what steers the individual effortlessly from one space into the next. At its best – when it's been well planned and implemented – a good flow within a place isn't something glaringly obvious but, rather, something that makes you feel comfortable and at home. When it hasn't, though, the likelihood is you'll feel limited within a space and there won't be a coherence between all the rooms as a whole."

In this apartment, attention to detail does much to enforce – albeit very subtly – the sense of flow. Exquisite flooring, dating from the 18th-century, restored by craftsmen and hand-laid block by block, seamlessly links the drawing room to the adjacent study-cum-library, for instance. Similarly, the hall and kitchen-diner are linked courtesy of a bespoke design floor that features Belgian hardstone edged with oak.

One of the secrets to a well-designed room is its ability to create

JEAN-MICHEL FRANK

an impact, and, as the first place any visitor sees, it is particularly important to make a bold impression within a hallway. Says Thong Lei of Decoration Empire, "You don't want anything bland in this part of a home – you need to say to the visitor, 'you've arrived - you're somewhere special'." In this case, the designers chose to make a statement by using a classic, wide, black-and-white striped wallpaper, punctuated by a striking portrait and a low black-silk-upholstered sofa.

A designer can't implement the decorative tastes and desires of the client unless he or she first listens to them, absorbs what they have to say, and – in many cases – is able to 'read between the lines', too. In this apartment, the result of that listening process is a home that both reflects and embraces the passions of the couple who live in it. A key area here, for instance, is a library-cum-study that allows the couple both to display and enjoy their extensive book collection. But of course it's not just about devising a practical space, it's also about creating atmosphere and paying keen attention to detail. Here, the mood is intimate and clubby, with hand-carved, finely detailed pine-panelled walls and bookshelves whose columns have been carved into 'folds', echoing the pages of a book. Adding to the feel are his-and-hers desks allowing the couple to work together but separately, a simple fireplace and capacious leather tub chairs. The result is a look that's redolent of historic English manor houses.

Atmosphere is key, too, to the success of the kitchen-dining room, where the challenge was to come up with a space that's as convivial to eat, relax and entertain in as it is practical to cook in. Here, bespoke cupboards – pieces of furniture in their own right rather than utilitarian units – have been painted to create a warm, relaxed French feel and maximise the natural light that floods into the space via two huge bay windows. Meanwhile, open shelving allows the owners' informal collection of pretty blue-and-white china, their cooking implements and glassware to become features of the room themselves. In the dining area, which merges seamlessly with the kitchen, wicker dining chairs and gingham cushions add to the cosy, countrified feel. In this room though, it's all about contrast. The light colour of the cupboards against the dark Belgian hardstone surfaces; on the table, silver combined with mix-and-match glassware and country-style cream tableware, so the feel is both rich and relaxed. The result is a cosy, elegant room that's both high-functioning and lovely to spend time in.

Throughout the apartment, colour has been vital in creating mood, from the warm pinky-terracotta and raspberry shades used in the drawing room that reflect the owners' love of relaxed Italian country living, to the serene Swedish blue of the master bedroom and the soft grey of the bathroom. Balance creates a sense of calm and harmony: it's particularly evident in the drawing room, where two sofas are complemented by twin chairs, while two stools in turn echo the apertures on either side of the fireplace that lead to the library-cum-study. It's also about throwing in the unexpected – that element of surprise that's key to making a space feel truly different. In this apartment, the cushions in the guest rooms are not covered in high-end silk but in inexpensive stripes and ticking, and are combined with rustic wrought-iron furniture to give these rooms a relaxed, easy-going feel. In the magnificent marbled bathroom, meanwhile, two arches, rescued from a dilapidated Italian theatre, have been reinvented to create a highly unusual open storage.

Above all, perhaps, this is a home that reflects its owners' sense of history, from family history via the collection of photos on display in the guest room and treasures that they've accumulated over their years together scattered throughout the apartment, to history in a wider sense, with their choice of furnishings and design details revealing a love of neo-classical-style pieces (the console tables and lamps in the drawing room, for instance), Gustavian furnishings and period finishes; the paint effect used on the cupboard doors in their bedroom, for example, was created using a technique that dates from the 16th century. It reflects, too, their passion for beauty combined with practicality and comfort. It's the sort of home where the period feel of the drawing room won't be sullied by the modern lines of a permanently-present TV, but where a motorised map above the fireplace lifts to reveal one for those nights when there's something unmissable on-screen. It's not a place that's about style over substance, then it's about style *and* substance. Inviting, beautiful and hospitable, this is an apartment that both its owners and their guests are delighted to spend time in, a place that most decidedly feels like home.

De Vrije Fries
je Fries
87 2007
85 2005
84 2004

Symmetry and balance are essential if a restful sense of harmony is to be created within a room.

NAGIEB MAHFOEZ Kinderen van Gabalawi
STRONG DEMOCRACY
PARTICIPATORY POLITICS FOR A NEW AGE
A STORM IN FLANDERS
WINSTON GROOM
BEL CANTO
WHY WOMEN MEAN BUSINESS
SUCCESS INTELLIGENCE Robert Holden, Ph.D.
THE FIFTH DISCIPLINE

GRIEKENLAND EN ROME
AMERICAN CIVILIZATION
MAXFIELD PARRISH
HOLLAND
STEINBECK
ANTHONY BURGESS
EARTHLY POWERS
WILLIAM BOYD
RESTLESS

CUTLERY

IV

The art of **Interpretation**

There is no more vital skill for an interior designer than the ability to listen to the client. As much as it's about hearing what the client says they want from their home, it's about what they don't say - about working together to build up a true picture of their tastes and passions.

Defining the skill of interpretation is relatively straight-forward – putting it into practice is considerably harder. At its simplest, interpretation in the interior design context involves listening very carefully to the client, and making sense of what it is the client tells you he or she wants from their home. But what if the client isn't actually sure what he or she wants? What about the vast range of shades and nuances between what a person might say and what they really mean? That's where any good designer needs a finely tuned antenna if he or she is to gauge what their client is truly after. And maybe they'll have to work, too, to understand their client's tastes, which may easily be open to misunderstanding. Says Thong Lei, "Often, I may have a client who says, 'I don't like green,' but you find them an object that they love in green, couple it with something else, and, suddenly, you find green's their favourite colour! Plus it may be possible, of course, that *their* idea of green isn't *my* idea of green, and vice versa."

Something that's essential in the first instance, before any project truly gets started, is an understanding of how clients actually use their home. For instance, if they're the sort of people who entertain for work a lot, and therefore require a formal dining room, it may make sense to place this in the darkest area of the apartment if they're only going to use

the room at night time. Alternatively, they may say that they love to feel the sun on their faces when they're sitting down to morning coffee, in which case a designer will need to create a suitable spot for doing so within any new layout. It's so important to concentrate on this kind of detail and to get it exactly right because, for all of us – not least someone who's gone to the trouble of hiring an interior designer – the home is the ultimate

sanctuary, the place where we ought to be able to relax and feel truly at one with ourselves and our surroundings. At Decoration Empire, the team will often begin by asking the clients for a 'wish list' of what they feel they can't live without in terms of room spaces and furnishings. Some may feel strongly that they want open-plan living, for instance, while others may see their home as more of a cocoon. There are practical considerations to be taken into account, too, plus the all-important need to manage a client's expectations. Thus, for instance, at the most prosaic level, budget is an absolute key area for discussion and interpretation – the client needs to know that the designer is aware of how much he wants to spend, and of course the designer needs to know what the client is willing to pay!

If a designer is to interpret his or her client's tastes successfully, there has to be a relationship, a rapport between the two parties.

"I guess it's no surprise that many of our clients have become personal friends" confirms Anne Noordam. At the level at which Decoration Empire works, the clientele in question are highly successful, sophisticated and cosmopolitan – while they know their own minds, one of the drivers behind their success is an openess to new suggestions. In the rare instances when clients are less sure of what they like, not only will a lengthy

process of discussion be involved, but possibly trips to art exhibitions and galleries, too. In this way, the designer can learn what it is that appeals to his or her client's taste, what it is that truly 'lights their fire'. This, of course, varies widely from person to person – some will respond passionately to colour, others to art or furniture, another to pattern – often it's a combination of several elements. Historical genres and the collection passions of the client are sure to play a role, too. For instance, if he or she has a special collection of art or – as in the case of the gentleman of the apartment covered in *Dramatic yet Calm* – weaponry, the designer needs to understand that this is significant enough to its owner to be worthy of a display area all to itself.

The more a client is set in his or her ways, the greater the challenge for the designer. Of course, it's great to work with clients who know what they like, but if they're unwilling to consider fresh combinations of colours or artefacts, it's going to be hard to convince them of something new and different. Moreover, the designer may be required to work around the client's stubbornness, rather than the two parties working together. Similarly, from a client's perspective, there's nothing worse than a designer who seeks to enforce his or her own taste, regardless of whether or not it suits the client. "We don't have a standard 'box of tricks' that we pull out on each and every project,"says Anne Noordam, "Instead, we're dependent upon the client's wishes--and that's why each and every one of our projects is as unique as the individuals the homes belong to."

Above all then, perhaps, the secret to interpreting a client's desire is trust – and the need to build a relationship that's based on its solid foundation. For it's only if a client feels the designer truly grasps what makes him or her 'tick' that he's likely to relax and respond to what that designer is suggesting by way of change. It's the 'ying' and the 'yang' between the two parties, a simpatico relationship of mutual feedback and reliance upon one another for ideas and inspiration. Once this relationship has been established, with the designer able to interpret his or her client's ideas and that client being confident in the designer's interpretation, that a project can move forwards once this stage is reached, anything is possible.

V

Penthouse | 484 m^2. Length of project: 13 months. Contents: three bedrooms, two bathrooms, dressing room, kitchen, informal dining room with fireplace, living room, formal dining room, two study rooms, laundry room and auxiliary room for kitchen.

Dramatic yet Calm

It's a well known interior design challenge: take two people with two very different sets of styles and tastes, then create one home that they'll both love living in. In this seventh-floor penthouse apartment, close to Amsterdam's Schiphol airport, the two styles concerned were his feel for tradition and drama and her passion for the coolly understated and the more contemporary. And it wasn't only the owners who had to be made to feel at home in the new apartment: his magnificent collection of 17th-century paintings, weaponry and armour needed to be accommodated sympathetically, too – no small task in a modern home.

The resulting interior proves that it is indeed possible to fuse two contrasting looks both seamlessly and beautifully. The secret has been to take the best from each style, blending them to ensure that richness and tradition never descend into extravagant stuffiness, while the understated is never reduced to the merely dull and cool never equals cold.

As ever with any home renovation, the most pressing need was to get the practicalities right – until this happens, it's impossible to plan any successful decorative scheme. In this modern building, two adjacent apartments were knocked into one, leaving Decoration Empire to plan a new layout for the resulting 484 m² of space that included two parallel hallways, three bedrooms and two bathrooms. Although every room makes a statement in its own right, it was decided that particular highlights within the apartment were to be the formal dining room, the kitchen (designed primarily with the lady of the home, a keen cook, in mind) and the gentleman's study. The hall, too, makes an immediate impact. Here, the geometry of graceful Corinthian columns, an inlaid floor and the unusual octagonal shape of the 17th-century portrait that provides the space's focal point is combined with the softness of atmospheric lighting, and each shade is specially lined in silk to cast a warm, cosy glow.

A particular challenge presented by the building itself was how to handle the rounded walls in some areas of the apartment. While these give the space a sense of momentum and flow, a lack of corners and edges to a room can of course pose practical problems when it comes to creating, for instance, cupboard space. By filling the apartment with specifically designed furniture where necessary, though, the design team have – in the master bathroom, for example – highlighted these curves as a softly undulating feature, using them to sweep the eye out and into the dressing room beyond, with its capacious built-in wardrobes. Here, the rounded walls are also emphasised by specially designed lampshades in a curved semi-circular shape that's similar to that of the space. It's just this sort of attention to detail that makes the difference between a home that's merely furnished, and a home that's truly finished; in this way, the designers have transformed the walls into an asset, rather than an awkward disadvantage. In the master bathroom, the painstaking attention to detail that typifies the project is particularly evident. The marble used here was personally inspected for quality by the design team, who flew one of its experts out to Italy to ensure that each slab would complement and enhance the room's design. Says Decoration Empire's Thong Lei, "Nature doesn't do uniformity, so it was up to us to make sure that the patterns on each piece would work well with one another." The practical, of course, is just as important here as the aesthetic and a useful touch has been the introduction of his and hers basins, always a time-saving addition in a bathroom.

Again on a more prosaic level, once the apartment had been plumbed and wired, a centralised, programmable Domotica system was installed. This means that lights can be switched on and off, curtains drawn, music played – even plants watered, both automatically and remotely. Moreover, the apartment's existing windows were replaced with heated ones. Says designer Anne Noordam, "This isn't about gimmicks or gadgetry for gadgetry's sake; it's about ensuring that a home can offer those who live within its walls the ultimate basic luxuries: total comfort and ease of living."

When it comes to the colour palette used in the apartment, soft shades have been key to establishing a relaxed tone and a sense of calm. In the drawing room, parchment coloured walls and curtains are teamed with blue-grey velvet-coloured sofas and chairs and hand-laid parquet flooring in bleached oak. Given the richness of the extraordinary Golden Age paintings hanging here, a darker wall colour might well have seemed oppressive. Instead, a neutral backdrop allows the artwork to dominate the room without overpowering it. The green of the velvet sofas, meanwhile, echoes the warmer green of the kitchen and informal living room beyond, creating a sense of continuity, and thus flow, from room to room. Stopping anything from looking too 'samey', too formulaic are flashes of sharp, unexpected and refreshing colour. This comes via a choice of accessories and ornaments, such as the raspberry-pink lampshade in the informal living room and the jade-green statuette on the mantelshelf in the drawing room. Also in the drawing room are two small, delicate Louis XVI wall-mounted consoles, an example of the advantages of design team and owners working closely together. The couple originally wanted these pieces placed one at either end of the room, but the design team were concerned that such a display would fail to make an impact within the space. Instead, it was agreed they'd be placed side by side. Now, with each holding a chandelier and a Dutch drawing of the 17th century hung above the two, a small vignette has been created and serves as a pretty feature within the room. It's a similar story in the formal dining room, which provides an ideal setting for antique pieces that include a 18th-century Dutch armoire and a little black and ivory cabinet dating from the 17th century from Germany. Display is all when it comes to works of art, and, to ensure that the cabinet is shown to its best advantage, an ebony coloured wooden plinth was created to hold

The Sketchbook of
Jan van Goyen

it. Paintings are displayed with equal care. In this room, they are hung low on the walls. In many homes, it's a common error to hang artwork too high, but a lower level not only allows the works to be viewed easily, it accentuates the height of the apartment's ceilings, too. The dining table is another example of practicality combining with beauty: an extendable mahogany piece was chosen so that the sociable owners can easily accommodate more guests around it when they're entertaining. On the dining chairs, the green and yellow shades of the rich damask silk upholstery cleverly play upon the contrasting colours of the wall and skirting board.

The challenge of using fine period pieces while at the same time keeping the overall feel of the rooms light and informal is brilliantly exemplified by the decorative scheme in the study. Here, pine panelling on the walls and a laid floor of French oak prevent the look from seeming overly traditional or fusty because of their pale colour and simple finishes. This in turn allows the collection of weaponry and armour – much of which has been wall-hung – to serve as a fabulous talking point, rather than an incongruity, within the space. It's a decidedly masculine room, a retreat for the gentleman of the house and, as such, the rich beige colour of the wood combined with the warm terracotta of the upholstery and blinds, plus a similar shade that's echoed in the coffee table (a piece made using an old Chinese screen) gives it an intimate, clubby feel.

But, less anyone should try to pigeon-hole and predict the tastes of the apartment's owners, it's worth pointing out the sheer variety of treasures they have given a home to. It's not just significant Dutch historical pieces that work well here – from paintings to the traditional many-spouted tulip vase on the drawing room coffee table – there's also room for Chinese porcelain pieces, from plates to ginger jars and statues, plus an ancient African mask on the little console table in the hall. With their simplified shapes and colours, these pieces have a decidedly contemporary feel that contrasts with their age and provenance.

In the kitchen, too, traditional and contemporary have been cleverly combined: while the bespoke kitchen cabinets and dressers – here painted a soft green – are classic in style, the overall message of this room is all about clean, contemporary lines: those used in the cabinets' design, the simple detailing of the bone doorknobs, the lines of the wooden shutter at the window and the graphic statement made by the extractor fan that hangs above the central granite-topped island. It's also visible on the floor, an angular pattern that features large blocks of Belgian hardstone, each hand-cut and hand laid, then surrounded by an oak inlay. The floor then carries through to the informal dining area, as does the soft green paint colour used on the kitchen cabinets. With its panelled Gustavian-style upholstered chairs pulled up close to the simple open fireplace, the dining room has a convivial, inviting atmosphere that's soothing in its lack of fuss. Adding panelling to the walls in this room not only creates visual interest, but a sense of history, too: who would guess, if they didn't know, that this room isn't one in an 18th-century country manor but in a 21st-century apartment. Above the little fireplace hangs a Pissaro, selected by the lady of the house from Decoration Empire's specially sourced collection. The team travels all over the world, to fairs, markets, galleries and shops in a bid to unearth treasures that they feel will appeal to its clientele, and this is just one example of the standard of art that the company is able to offer its clients.

Again, the difference in this apartment is in the detail: in the master bedroom, rich shades never become overly opulent. Rather, they're kept in check because they're expressed in simple forms – witness the 1950s' emerald glass lamp on the bedside table with its plain silk shade and the unfussy finishing of the bed's velvet-covered headboard. Although the apartment is graced with many museum-quality pieces, what makes it special is that its design never forgets that true style has little to do with cost: it's about comfort, consideration and beauty. In the guest bathroom for instance, warm colour and detailing comes courtesy of the tiles on the floor – inexpensive ones that have been produced and used in Dutch factories and shops for centuries. Here, teamed with marble-topped French-style painted vanity units and a traditional roll-top bath, and with the grey of the tiles echoed on the walls, they take on a whole new identity.

With each half of the owner equation feeling that the apartment now meets their practical needs and perfectly reflects their tastes and passions, then, this is a home where traditional charm meets contemporary cool. The result is a space where comfort and beauty sit side by side, a harmonious marriage of ease and luxury.

The key elements
of proportion, colour,
pattern and light
here combined to lead
the eye to a beautiful
focal point.

MUSEUM BREDIUS
MUSEUM BREDIUS
ERFGOED VAN NEDERLAND

VI

The art of **Flow**

It's flow that makes a collection of rooms feel like a harmonious, 'seamless' home, that helps to propel visitors gently from room to room. Yet, at its best, the individual is totally unaware that it's in action. To create flow requires an extremely subtle mix of the practical, the aesthetic and the instinctive.

Flow is arguably the most elusive element within any successful home design. At its simplest, it's the unseen, unobtrusive energy that gently guides one from room to room, and that succeeds in uniting all those rooms so that nothing jars, everything fits together. Get the flow in a home right and, as its name suggests, the spaces will seem to flow effortlessly, one into the other, to create a harmonious whole. It sounds easy enough, but take a range of rooms of different shapes and sizes and with different styles – one, say, with a stainless-steel, industrial-looking kitchen, another a 1940s'-style living room, another a formal dining room – and, suddenly, the ability to somehow connect all these spaces, to make them look as if they do all belong together, becomes rather more complicated. That's where the work of an experienced interior designer really comes into play. Says Thong Lei, "Because the issue of flow is potentially so challenging, it can push many people into 'playing safe'. After all, it's a lot easier to create a coherent-looking home if all the rooms are painted in similar colours, or dressed in similar styles. But that makes for a very dull feel to a place – and few things are more energy-sapping than boredom. That's why we delight in putting different styles and colours together and making it work. It's refreshing for everyone: for those who work on a project, our clients and of course for the home itself."

Again, the ability to mix and merge styles can be learned through experience – but only to a point. Ask any great interior design team and they'll confirm it has more to do with instinct than it does with standard rules. And besides, many of the factors that make for great flow seem in direct contradiction to any pre-conceived rules. As Anne Noordam points out, "There was a common belief for a while that, to achieve ultimate calm and flow within a house, minimalism was the way ahead. We've found, though, that nothing is more restful, more harmonious than a collection of

objects in a room that all work well together." Perhaps in some ways, this is down to the fact that homes, of course, are all about the people who live within them, and just as humans tend to work better and feel happier in communities than they do in isolation, so do artefacts. In other words, they reflect the emotions of the people they belong to, and it is that that imbues them with their own hard-to-pin-down 'soul'.

Beyond the philosophical, however, there are practical issues that undoubtedly promote flow within a home. Within an apartment, which is often based on one floor only and where space is likely to be more limited than it would be within a house, it's particularly important to ensure spaces are somehow linked together. One way to achieve this is to open out areas, adding door-free apertures between different spaces; to separate a living and dining room, for instance, but without cutting them off entirely from one another. The effect is cosier than the 'open-plan living' approach (which can also be impractical, particularly if there are several people, all doing different things at different times living within a space) but still maintains a sense of airiness and inter-connection. Another practical, subtle way to link rooms for a sense of flow isn't by wall covering or colour or fabric, but via the floor. Because we tend to look up rather than down, a beautiful floor will never dominate a space, but can be used to provide continuity if it's used throughout an apartment. Symmetry, too, plays a part in conveying a sense of flow – a pair of lamps, a pair of tables carefully arranged on either

side of a fireplace, with another pairing of items in a room beyond, help to unite the two different areas. Similarly, when it comes to colour, it's often not necessary to unite great swathes of wall with similar shades; particularly if they're bold primary colours such as yellow or red. Rather, it's a case of planting visual 'reminders' of those colours in each room: a fragment of yellow wallpaper can be echoed in another room by, for example, a set of dining chairs upholstered in yellow. By keeping such links understated, the effect remains subtle and the look doesn't jar. It also means that connections can be made within rooms of different styles, ensuring that they always share a common factor, even if that common factor only plays a small part in their design as a whole. Finally – but vitally – the energy that any painting or sculpture resonates is invaluable in generating a momentum of its own, a momentum that helps push one space into the next, moving the inhabitants on both emotionally and physically.

Flow, then, is what makes a home feel comfortable. No amount of fabulous art, gorgeous furnishings or great architecture will make people feel truly at ease in a home if it doesn't have it. The secret to creating it, it seems, is almost as elusive as a definition of the word itself. Yes, it's about using elements of the design process sympathetically and appropriately, but it's more than that. For a truly original interior designer, it's about having the instinct to find or make connections between elements and areas that others might easily have overlooked, or that would never have occurred to them in the first place. It's the sort of gift that only goes to support the maxim that the best interior designers are born, not made.

VII

Art lovers' apartment | 215 m^2. Length of project: 13 months.

Contents: two bedrooms, two bathrooms, dressing room, living room, dining area, kitchen, two large balconies.

Rooms with a View

This apartment is all about colour, art and energy. Much of its sense of momentum initially comes via its spectacular location: seven storeys up in a converted former office block, with fantastic views across one of Belgium's loveliest and most historic cities. Because Decoration Empire had already designed a home in the south of France for the dynamic and high-flying couple (he's a property developer, she's an art historian), the team was already aware of their love of modern art and their cosmopolitan lifestyle. Their Belgian home would reflect their personalities, tastes and passions, but this time with a different twist – and a heightened sense of drama.

Because the apartment's hallmark is the bank of glass windows on the furthest wall offering panoramic views over the city, light was never going to be an issue here. Northern European light, of course, is different from that of the Mediterranean: it's sharper and cooler but exceptionally clear. This meant that no one involved in the project had to worry about the effect of strong, dark colour on the spaces – the apartment could take deeper shades without the risk of it appearing gloomy – and flashes of sharp and rich primary colours, such as red and yellow, have been added for interest and contrast.

What's particularly striking in this home is the way in which openness contrasts with a sense of mystery, of secret spaces. Its open-plan layout – with the bedroom just round the corner from the living area and without a door to separate the two from one another – undoubtedly adds to the sense of airiness and flow. In the hallway area, however, the door leading out of the apartment is unobtrusive, almost hidden within a black-lacquered, bespoke wooden panel, just one of many that covers the wall on this side of the space. In effect, then, it appears almost to be a secret door, one that a visitor might stumble upon by accident. The sense of surprise is further heightened by the decor of the hallway itself, with its antique-mirror-lined walls consisting of panels specially made by Dutch-based craftsmen and edged with brass. What's particularly beautiful is the foxed finish to the glass, which gives it a smokey, dappled appearance. And the mirror here is used for more than mere visual effect: in practical terms it allows light to bounce around the apartment, increases the sense of space and creates a wealth of interesting vistas, with views reflected within views. Finally, as if there wasn't enough momentum in this space, a sculpture in plaster by the artist Laszlo Szabo, displayed on its own specially-designed plinth acts as a visual anchor, resonating its own positive energy and giving the hall an unexpected sense of serenity and calm.

A sense of coherence goes hand-in-hand with a sense of flow, and, in this apartment, choice of flooring does much to underpin it – albeit it in the subtlest of ways. The hand-cut and hand-laid oak mosaic parquet ages beautifully, taking on the soft, mottled greys of the apartment's walls and linking the various spaces. Its design is echoed further on in this home, where the bespoke walnut dining table – rounded to make group conversation as easy as possible – is also made up of a mosaic of wooden pieces. Around the table are chairs upholstered in zingy yellow fabric, again designed to be as comfortable as possible and making the table an enticing spot around which to chat or simply admire the views over the city. The yellow colour is picked up nearby, in a sleeping area, where a tightly-pattered wallpaper provides a warm backdrop for a magnificent black and gilded mirror and an unusual golden-coloured wooden table from China.

If ever an apartment exemplified the need for any well-designed space to have visual points that capture and hold the interest it's this one. From every angle, there's something arresting to look at, a vignette to surprise or delight. Each of these, though works well within the space as a whole, never serving as a

distraction but only adding to the sense of harmony. One of the best examples sits quietly but dramatically by the hidden front door: a marble-topped carved-wooden table from France, dating from the 18th century, becomes host to an old lion's head door-knocker and a neo-classical candelabra that's been transformed into a lamp. What brings the whole vignette to life, however, is the extraordinary rich-red gessoed work by Austrian artist Jacob Gasteiger. It brings to this little area three essential things: drama, a sense of scale and – because of the pleated-look of the paint – a sense of texture. Moreover, it's cleverly 'reflected' on the other side of the room but in larger dimensions to fill the larger area. Another carved, marble-topped table – this time from Ireland – is also topped with lamps and offset by a red and white collage, by Jean Paul Pincemin. The effect of this subtle visual repetition isn't only striking to look at, it creates a sense of balance and thus of harmony, too. That's what stops such dramatic, colourful pieces from looking over-powering – instead, and in combination with one another, their effect manages to be both restful and dynamic.

This apartment exudes a quiet, understated sense of luxury – not only in the many beautiful objects it provides a setting for, but in the materials used in its design and the quality of finish. Take the walls of the bedroom area, for instance: they're covered in padded panels that are upholstered in soft turquoise fabric woven from finest horsehair for a unique texture and finish. Making a graphic statement here is a simple black stained arched table, from the Decoration Empire Home collection. Meanwhile, in the master bathroom, detail in the marbled floor is provided by an arc of mosaic tiles in platinum – one that extends all the way out and up onto the shower area wall beyond. In the shower area, too the *Fior di Pesco* marble on the walls is remarkable in its detailing and colour. The choice to use exquisite materials such as horsehair fabric, lacquer, marble and platinum is based on a desire to create the best possible impact – not in any ostentatious way, but in a way that reflects the owners' commitment to quality and craftsmanship, their unwillingness to compromise. Creating continuity, the mirrored theme first seen in the hallway continues in the bathroom, with its 1940s'-style furniture, and in the magnificent, unashamedly glamorous dressing room – where, again, bespoke antique-mirrored panels have been used to create wall-to-wall cupboards, offset by brass insets with their own warm, mottled patina.

Rooms that need to be practical as well as beautiful fulfil their brief, too: the kitchen combines sleek, contemporary stainless-steel cabinets – made to Decoration Empire's design – with natural materials: granite for the work surface, the same mosaic-effect oak on the floor. These natural materials succeed in softening the feel of the room and preventing it from looking too industrial.

In this home, above anything else, it's art that's the star. A deceptively simple-looking dark paint effect on the wall (though all one colour, it takes around six coats to achieve its varigated finish) provides an ideal backdrop for the beautiful art collection of the owners. Each of these plays a vital role in creating mood within the apartment, and – whether large or small – creates a sense of scale and makes a definite statement. These are all monolithic pieces, from the vast ab–stract canvas by François Millon that overlooks a gilded Italian table of the 18th-century and pair of canvases by the Dutch artist Leijerzapf behind the grand piano (the gentleman of the place is a talented pianist), to the work of painter Permaeke and the small trio of marble objects dating from the third millennium, stunning in the lyrical simplicity of their form. A sense of history is created here – one that reflects the owners' energy and tastes, their love of travel and their appreciation of quality. Says Decoration Empire's Thong Lei, "In a home like this, it's a case of less is more, in that the place never looks cluttered, there isn't that much stuff. What there is, though, is of exceptional quality, with every piece reinforcing the artistic value of every other piece. Adding new items to the mix helps to build up a look organically, layer on layer, so that nothing here looks forced or hastily chosen. Instead, this looks like a home that's been put together with real love and care – which is, of course, exactly the case."

PAST IN PRESENT
TERRA
Permeke
ITALIAN HIDEAWAYS
TRENDS AND TRADITIONS
RAYNER ■ SCHEZEN
NUDES

In any room you enter,
there should always be one object
or vignette that makes
a statement.

VIII

The art of **Colour**

Nothing provokes stronger reactions in a home than colour – and no two people respond to it in exactly the same way. How is it possible, then, to use a variety of colours within a home without creating disharmony? The answer lies in having both a confident eye and a steady nerve.

Nothing denotes the mood and tone of a home better than colour. But in apartments, using it can be difficult. Often, the temptation is to play it safe and to stick to blander colours, thinking that this will succeed in unifying a smaller home. It's worth following the decorative motto of the owner of the apartment featured in *Classic Meets Unconventional*, who chose to punctuate darkest grey with flashes of acidic yellow within her own study: 'No guts, no glory'. Because colour has been used here with confidence – utter conviction, it works and the result is that a colour that might not spring immediately to mind as a home-decor choice has a wonderfully refreshing, energising effect within a room. As Thong Lei puts it, "Space is a state of mind, and while people often think that dark colours, for example, will make a space look smaller, it's not always the case. And it's better, surely, to have a small space that looks 'right', that looks cosy and inviting or dramatic, than a larger-looking space that adds nothing to the feel of an apartment as a whole. It isn't only a lack of square metres that makes a space seem smaller, it's lack of thought and imagination."

Running hand-in-hand with colour is the concept of flow, and it's easy to think that a sense of the latter is dependent on the former. Not so. Flow can be achieved through a variety of means: by using similar flooring throughout an apartment, for instance, by opening up the spaces themselves or by echoing fragments of pattern and colour very subtly between areas of an apartment. If these sorts of techniques are used, it is indeed possible to create a home with a great sense of inter-connectedness and unity without any need to stick to the same colours throughout its rooms.

Decoration Empire's work undoubtedly proves that it's possible to use a wide range of different colours within one apartment, and still succeed in pulling the look together. This is particularly evident in Thong and Anne's own home, where the bluey-black of the dining room walls contrasts with the palest of greens in the living room. The team chooses every colour

specifically to suit the mood they want to create within a room, to energise it in a certain way: darker shades that will be offset by candlelight for night rooms, and paler shades to make the most of natural daylight, and, in the case of Thong and Anne's living room, to echo the green of the garden beyond. And of course different colours will have different, individual emotional resonances for each person. In most cases, though, even when people are adamant that they 'can't stand' a certain colour, they have only one specific shade of that colour in mind; they're almost sure to respond positively to it if it's used judiciously, in the right shade, or in careful combination with other colours.

Sometimes, too, colour is used to evoke a certain historical period or feel. Thus for many people, orange in combination with olive green or solid purple will always say '1970s', pale pinks and silver suggest the glamour of the 1930s, while rich, opulent burgundies and greens are redolent of the 17th-century royal courts of Europe.

When paint is used on walls or wood, a variety of techniques is available that will affect the final look of any colour. For instance, Decoration Empire uses a technique, visible in the living room shown in *Rooms with a View* where a patina of colour is built up over six layers, creating a very subtle, deliberately streaked effect, one that creates a wholly different mood from that which

a bank of solid colour would provide. Adding texture, too (the company sometimes mixes sand with its paints for a rougher finish) alters the final look and depth of a colour, as, of course, does natural light as it travels around a room during the course of the day. But colour need not only come via wall paint. In many of the most successfully designed rooms, it may not be the walls that stand out at all. Instead, they'll serve as a back-drop for coloured accessories, which will have their own sorts of colour depending on the material they're made from. Thus a porcelain Chinese vase in yellow will have a rich, lustrous colour, while a slubby linen's look will be more subdued. Often, too, it's the colour of an accessory that serves as the starting point for the design of an entire room: the orange in a painting, for example, being echoed not on walls, perhaps (unless it's a fragment of wall dressed with, say, an unusual wallpaper), but in flashes of upholstery on cushions or chair covers, or in a carefully edited collection of coloured glass.

Ultimately, though, as Anne Noordam confirms, there is no one palette of colours that a good designer would or should turn to. "For us, choice of colour within a room is about gut feeling, about instinct. We'd compare how we use colour to how a chef places food upon a plate. The 'wow factor' comes, at a restaurant, from the unexpected combination of flavours that truly delights the palete. With interior design, it comes from a combination of colours that delights the eye – and

the soul. As a viewer, you may not be entirely sure why it works; all you do know for sure is that it does. And that's what makes you feel completely comfortable, completely 'at home'."

IX

Dynamic apartment | 320 m². Length of project: 6 months. Contents: two bedrooms, two bathrooms, storage room, kitchen plus adjoining dining room, living room with fireplace, reception area with wall-to-wall bookcases, two study rooms.

Classic meets Unconventional

This apartment proves it's possible to create two entirely different styles of home within the same building – even when the same design team is involved. It's situated in the same modern mock-Tudor building as the apartment featured in 'Warmth and Light', yet both homes have a feel that's entirely their own. In the case of this apartment, the owners first fell in love with Decoration Empire's work when they toured a show home within the apartment block that the team had designed for its developers. The company was also responsible both for the restrained and elegant communal entrance hall in the building, and the decoration and furnishing of the quietly luxurious swimming-pool area.

Because the layout of the apartment had already been predetermined, the designers were largely concerned with its decoration and furnishing. However, they were keen to open up spaces wherever possible, fully exploiting the building's high ceilings and the light that floods into each room of this ground-floor apartment. In the wall separating the reception space from the living area, for instance, a couple of doorless apertures have been added to link the spaces without detracting from the sense of airiness. Columns in the Corinthian style, meanwhile, have been added to create subtle visual interest, their design having been pared down until they consist of little more than a few well-defined lines. In removing all extraneous detail, then, a traditional architectural device has been given a wholly contemporary twist, adding a sense of quirkiness to the space. Arguably, this area of the apartment best demonstrates the seamless fusion of contemporary with traditional feel: while the living room is pared down and decorated in striking black and white, with a minimalistic fireplace and dramatic black-painted inset shelves, the room beyond features curvaceous French chairs and full, period-style curtains in a traditional print. Connecting the two spaces are a lamp topped with a black shade in the study, the black echoing the colour that's predominant in the living room

and, adding a tiny, subtle detail, the black piping on the study-curtain tie backs. Finally, a wholly unexpected element within the mix takes the form of a funky zebra-print-covered ottoman. A conventional piece of furniture, then, has here been upholstered in an unpredictable style. Says Thong Lei, "It's actually extremely hard to add a surprise element to a room – get it wrong, and the result will look contrived. Get it right, though, and it can provide that all-important focus to a room that elevates it from the merely lovely to the really special."

Though the apartment's colour scheme is primarily one of neutrals such as cream, white and black, occasional use of acidic colour is another contemporary feature here; it has been applied judiciously to give certain rooms a delightfully off-beat, unconventional feel. In the kitchen, for instance, chairs have been upholstered using simple slipcovers in a linnen cotton mix, but their bright orange colour picks up on the flowers within the artwork on the wall and stops the darker pieces of furniture from seeming too sombre.

While the gentleman of this home is classic in his tastes, his wife's attitude tends to be one of 'no guts, no glory'. His study, then, features hand-made mahogany bookcases and an antique desk and armoire, plus cream walls and a pale oak floor teamed with Roman blinds in masculine black linen. In hers, meanwhile, she has had the courage to partner deep, warm grey with zingy Chinese yellow. The brighter shade never becomes overpowering, however, confined as it is to the USM steel bookshelves, the artwork – abstract pieces including a impasto piece by the Dutch artist Wil Jansen and a pair of works by the Spanish artist Eduardo Barco – and the striking pair of oversized Chinese vases. Furniture here, too, is as unexpected as the colour, with a corrugated paper Wiggle chair by Frank Gehry echoing the curvaceous shape of the vases, and its natural colour toning well with the hand-laid oak floor. Another interesting element of this room is the way in which pairings – of the vases and the artwork – are used to create a sense of harmony and balance. "We find that things often look better, feel more serene, when grouped together in twos or threes. It looks better than when they're displayed on their own."

Key to the success of this home is its sense of energy and movement, something that choice of art has done much to convey. In the hall, for instance, impact is immediately created by a modern coral-like sculpture that contrasts with the elegant 1940s' bronze and marble octagonal table on which it sits. Spotted in the window of a municipal gallery near Cannes, this sculpture was eventually purchased after its creator finally agreed to part with it. "We knew it would be perfect in this space," says Decoration Empire's Thong Lei, "It resonates energy – it's as if it's a life-force in its own right." The Cambodian piece on a side-table and collection of 2nd century AD Han Dynasty earthenware, both in the living room, have a similar effect; the timeless simplicity of their shape, the beauty of their finish and their quiet energy immediately appealed to the apartment's owners. In the comfortable master bedroom, too, an African sculpture on an antique chest of drawers is teamed with a black and white vase, specially made in Thailand to Decoration Empire's design and decorated with a pattern inspired by fishing nets. Although many of the objects in the apartment are new to the space and were specially chosen for it by the owners, a vital aspect of any home-design process – and sometimes a tricky one – is the incorporation of pre-existing treasures and possessions. Here, though, they have been effortlessly introduced: the little black and white pictures in the living room, the cream portrait in the bathroom and the bright canvas in the kitchen have all been used as starting points or inspiration for design choices regarding form and colour. One of the secrets to making art look as if it belongs in a new space, even if it's not at first glance a natural 'fit' within the rooms, is to have it remounted and reframed. This is just one of many of Decoration Empire's areas of expertise. Says Anne Noordam, who began her design career working exclusively within this field, "It's extraordinary just what a difference a frame can make to the look and feel of a work. In, say, a contemporary home, a traditional work, once reframed, will suddenly work within a space when before it might have seemed to jar with the rest of the pieces within a room."

In this home, then, it's all about contrast, about mixing simple, graphic shapes with more traditional and ornate furnishings, and about throwing in a liberal dose of the unorthodox. That's why you'll find the eccentric zebra-print upholstered ottoman in the otherwise restrained study, and a robust, curlicued wrought-iron chandelier – as opposed to the more traditional crystal light fitting – hanging over the table in the dining area. Also in this part of the house, a traditional mahogany table is teamed with generously padded and upholstered chairs. The correct dining chair is one of the elements that occurs in any home on which Decoration Empire has worked. So many dining chairs are elegant to look at but extremely uncomfortable to sit on, but the company ensures its clients select models that are both beautiful and as well-designed and relaxing to sit in as a living room chair. It's a small but essential touch, and one that helps to make meal times all the more comfortable and enjoyable.

As ever, it's an attention to detail and commitment to quality that marks out this home as being truly special. It wouldn't be an exaggeration to describe the lamp shades here, for example, as the haute couture of the light-fitting world. Each has been handmade in Paris using natural materials such as silk and linen. In many cases, they have been lined with a different fabric for a subtle, glowing effect. An excellent example of the way the traditional and contemporary are combined with practicality is shown by the positioning of such lights in the living room. While the shades on the bookcase-mounted lamps are traditional, they're positioned in a wholly fresh way, not at the top of the bookshelves but halfway down. This immediately has the effect of seeming to lower the apartment s high ceilings, making the space feel more intimate and cosy. Meanwhile, on the couple's bed, cushions have been upholstered in the finest fabrics – the zebra-skin-effect material, for instance, is a silk-velvet mix that's extraordinarily soft to the touch and that has been handpainted for a unique finish.

When it comes to delighting the senses, the guest bathroom is another space that more than fulfils expectations. On the mod-cons front, it boasts a state-of-the-art jacuzzi surrounded by Italian marble, while, in aesthetic terms, everything here has been selected to soothe and relax. The restful taupe colour scheme is complemented by a comfortable armchair from Decoration Empire's own collection, while the painting on the wall and a small display of 19th-century shells and fossils provides visual interest.

Sense of flow is conveyed in this home, then, by a one-off mix of ingredients: both the energy of the art and the directional herring-bone-effect bleached oak parquet flooring seem to direct visitors from room to room, while a subtle colour palette spiced up with flashes of colour and use of pattern provide connections between areas and features within the apartment. The wallpaper in the hall, for instance, is echoed in the dining area, where a similar pattern is repeated in larger form on the chairs' upholstery and then on the gentleman's study curtains.

THE HOAX
THE ROCK
SWEET CAROLINE
THE KENNEDY CURSE
Henry Kissinger
THE AMERICAN DREAM
SPENGLER
Camilla Läckberg

Mustard Pommery

DONNA TARTT
De VERBORGEN GESCHIEDENIS
Adriaan Venema
Ger Verrips

SOCRATES

A room should
have passion
and a soul

BORGEN
HIEDENIS
andere ogen

X

The art of **Scale**

Choosing pieces that work within the dimensions of a home – particularly an apartment – is far less straight forward than it sounds. Furnishings don't just have to 'fit' in practical terms, they must 'fit' in metaphorical ones, too – enhancing the area and creating illusions of space where necessary.

Every home needs airiness and space if its inhabitants are to feel they can breathe and relax within its walls. And, within an apartment, where space is often at a premium, it's especially important to create this sort of atmosphere. Scaling proportions within a home is in some ways one of the most practical aspects of interior design. It involves choosing pieces whose sizes complement one another and make their setting look either larger and grander or more intimate – or both – depending on the client's wishes and the style of the room. But it's also a hard-to-acquire skill, one that comes after years of experience and trial and error. Subtleties are involved: choose an 18th- century armoire for a smaller space and get the size right, and this bulky piece of furniture will actually succeed in making a space feel larger; get it wrong by choosing a piece that's a fraction too large and that same space will feel cramped and oppressive. Plus of course, one must always be realistic: where space is limited, even when it comes to room planning, there's no point in losing valuable square metreage to furniture that won't be fully utilised or appreciated. Not every home will be large enough for, say, a grand piano, and what point would there be in having one for effect, if one never actually plays it? Says Thong Lei, "The art of scale involves far more than choosing and placing existing pieces of furniture to achieve pleasing proportions. We will always look at the scale of everything in any home we tackle, and we've revised door heights, and replaced cornices, windows, skirting boards – everything that will alter the sense of space within a room until the proportions we're after are achieved. Lighting, too, plays a vital role, not just in creating atmosphere but ensuring spaces look larger or smaller than they may actually be – proportions can be emphasised and accentuated as the designer chooses, for a flattering airy or intimate effect."

To some extent, then, the art of creating an ideal sense of scale definitely has its hands-on aspects. But, as the team at Decoration Empire would confirm, it's also about having a certain sort of courage, a courage that influences how individuals will experience the rooms.

In smaller areas, such as one will often find in apartments, it's easy to be lulled into a sense that 'less is more', keeping furnishings to a minimum. Thong Lei again, "We'll often find, however, that 'more is more'." A beautiful, generous mix of carefully-chosen objects and pieces of furniture is the antithesis of mean – and rooms that show their custodians are generous with their tastes and passions will always feel larger than soulless spaces with little in them. That said, however, it certainly isn't a question of merely filling a room with 'things' – if they don't have meaning for the owner, haven't been selected with the utmost care, no amount of priceless objects will ever make a room 'feel' right. What is possible within a room is to create a sense of scale that focuses on one simple vignette. Take a table topped with little bronzes and a neoclassical lamp, then add a graceful Louis XVI chair nearby, and the tone and proportions of the whole room will be set. Where the process moves from the practical to the emotional, however, is where a designer's instinct kicks in – that ability to throw something into the mix that's totally out of proportion to everything else within the room, yet somehow serves to emphasise that room's sense of scale. "Yes, this is a skill that certainly can't be taught," confirms Anne Noordam, "Thong, for instance, has the ability to be taken into a room blindfolded and he can still sense its proportions. He has a very instinctive, finely-tuned sense of spatial awareness."

Proportion and scale, arguably more than any other aspects of interior design, can affect the atmosphere of a room. A space that's filled with colour-toning pieces – with curtains and upholstery matching the colour of a carpet – or furniture of a similar style is always going to feel more restful than a room in which there's conflict. Such a room, though, will always run the risk of appearing too bland and without impact – after all, it doesn't take much for a room to move from being 'restful' to 'soporific' – and that's an atmosphere, surely, that one has no desire to create anywhere bar the bedroom! Again, this is when it takes the experienced eye of an interior designer to introduce elements of surprise that prevent the room from looking too 'safe' and feeling dull.

As with so many aspects of interior design, the yin and yang of the process fuses the practical with the emotional. While there are practical steps anyone can take to increase or decrease the sense of proportion and scale within a room – anything from lowering a ceiling or painting it darker to give it a sense of having been lowered, to raising the position of curtain poles and door frames to make ceilings appear higher – it takes an elusive talent to make a room *feel* larger or smaller. Says Thong Lei, "Nothing scales up a home more than its having an atmosphere that's warm-hearted and welcoming. Homes with a cold atmosphere – no matter how grand in scale – will always feel cramped, if only in emotional terms. But it's those emotional terms, far more than practical ones, that truly influence how one responds to a space. Generously hospitable homes will always, somehow, succeed in feeling generously proportioned."

XI

Designers' apartment | 480 m^2. Length of project: ongoing! Contents: living room, library, study, dining room, bedroom, dressing room, bathroom, small second bathroom, fitness annexe, laundry room.

Labour of Love

Anne Noordam and Thong Lei's own home in Gouda is a testament to the passion and care they put into all their design projects, including their own. When they bought it 15 years ago, they acquired just the bare bones of one large space, situated in what in previous incarnations, had been first an old *stroopwafel* factory (a stroopwafel, for the uninitiated, is an irresistible syrup-filled wafer biscuit, a Dutch favourite with very good reason), then an earthenware manufacturer's. Although the place, which was built in 1917, was clearly going to require a lot of work to turn it into their ideal home, the couple could immediately see the potential of its 3.6 metre-high ceilings and period-industrial feel, plus there were still some original features that made the place special, such as roof beams and rafters. These only required cleaning and varnishing for a new lease of life. Other essential jobs that had to be tackled before they could really start to furnish the space included knocking down the few existing walls, that made certain areas feel pokey, and rewiring and replumbing the entire apartment.

Because it's situated at ground-floor level without an abundance of windows, the space wasn't particularly light, so another of Anne and Thong's first jobs was to open out the vaulted kitchen roof, plastering just so far up the walls, then leaving the original structure with its warm-looking distressed finish exposed. Finally, the roof, which they had re-glazed, allows the light to flood in. They also took care to plan their room layout according to the available natural light within the home, so that their blue-black painted dining room – which of course is used primarily in darkness anyway – is situated in the least light area of the house, while the living room overlooks the garden and is thus well positioned to capitalise on any sunlight during the day. Echoing the greenery in the space beyond the apartment, Anne and Thong chose to paint the room's walls in a calming *eau de nil* with toning matte-silk curtains. Beyond, in the library, this colour is repeated in a darker shade, creating a sense of continuity from room to room.

All of Thong and Anne's own design tenets come into play within their own home. For a start, this is somewhere that allows them to indulge in their own particular hobbies and interests. Says Thong,

"What worked well was the way in which, without really planning to do so, we each took responsibility for particular spaces. Anne went for the library and the bathroom because she enjoys relaxed, quietly luxurious spaces where she can read or unwind, and I designed the kitchen because I'm a really keen cook."

Here, Thong has created a room that shows he really means business when it comes to mealtimes. This is a working room, with two ovens, a vast sink and a state-of-the art La Cornue range. Specially designed stainless steel units, topped with hard-wearing granite, allow for plenty of practical, no-nonsense storage, with a heated Belgian hardstone floor underfoot. The glazed ceiling makes for clear light in the room, ensuring that Thong can see exactly what he's doing while he works – pretty essential given that, on a regular basis, he'll cook for as many as 30 friends. As with any Decoration Empire home, however, there's always room for art, and the space above the sink is dominated by a striking, circular polyurethane work in red, made as a gift to the couple by a previous employee. "It's full of energy yet also seems to exude a Zen-like calm – just the sort of spirit one needs in the kitchen," comments Thong.

Other spaces in the apartment show just how important colour and form can be as the starting point for the complete look of a room. In the couple's bedroom, for instance, Anne took inspiration from a beloved Italian bureau in a rich raspberry colour, which dates from the 18th century, is decorated in the Chinoiserie style, and which she bought from an Italian antique dealer in Paris. "We wanted a feel here that reflects that sort of look: dramatic, sophisticated, a Venetian feel," she says. They've undoubtedly achieved it by mixing, in their unique style, layer upon layer of sumptuous fabrics, papers and patterns. The headboard alone features a hand-carved and gilded frame made by Decoration Empire's craftsmen, with silk gauffrage upholstery in raspberry. The heavy silk curtains combine deep green and, in their silk lining, old pink to romantic effect, tumbling down as they do from the antique wooden canopy above the bed. The theatrical look is offset by the hand-blocked Zuber wallpaper. Despite its flamboyant pattern, the paper doesn't dominate but merely complements the rest of the space, because the grand scale of the room can take this sort of exuberance. And a small art work near the bed illustrates Anne's expertise in framing: a relatively inexpensive sketch is surrounded by a hand-carved frame of exceptional quality, the two partnering one another perfectly. Finally, for a sense of balance, the floor is carpeted with a beautiful hand-woven French design that's exceptionally soft and warm underfoot. "This is a room to escape into, an ultra-relaxing, romantic sanctuary," says Anne, "As a space, it's ended up exactly as we first envisaged it. After all, you spend an average of eight hours a day in the bedroom, so you might as well ensure it's somewhere that truly delights you."

Anne and Thong's skill as collectors is well documented, and their home is an eclectic mix of treasures that they've fallen in love with over the years. Sometimes, they've seen something, haven't had a place for it at the time, but have hung onto it, knowing that one day they would find the perfect use for it. Such was the case with the mercury glass doors in the dressing room, which, with their mottled patina, make an intriguing statement and reflect and diffuse light in a magical way. The glass, which is thought to date from the 18th century, was rescued from a ballroom of a grand house. It was kept in storage for some four years, until Thong and Anne had the matte lacquered frames made to work around them, and installed them in their own home. Another outstanding salvage feature is the intricate herringbone-patterned oak floor that dates from the 19th century and was once underfoot at the Museum of Natural History in Paris.

Then there are the bizarre pieces that somehow work together so well: a green Lalique vase combined with an old death mask, a piece of fossilised coral and a wooden fertility symbol from Africa, all displayed on top of an industrial table that dates from the 1960s. The combination of old and new, of different colours and scales of objects can't fail to create an impact; but, above all, it's the sense of passion that unifies this collection and that makes for such an eye-catching vignette. There are pieces relating to Thong's Chinese heritage, too, such as the calligraphy screen in the living room and the oversized pair of Chinese blue and white vases, and elsewhere, a little photo of his grandfather, who lived to be 102. "No home can possibly have a soul," he says, "unless it's a home, not just for people, but for well-loved pieces with a sense of history about them."

Something else that makes these kinds of collections so special is the fact that they're indefinable: they're not drawn from any one artistic genre or historic period: a 1960s' French Wiggle chair can sit comfortably next to a classic silk-velvet-upholstered sofa,

Sotheby's
PERRIN
Influential Interiors
Maisons du Quercy et du Périgord
GA HOUSES 52 PROJECT 1997

MATERIAL JOURNEYS
MIDAS DEKKERS
POOT
JOSEPH O'NEILL
LAAGLAND
FANG
LUBA
BUDDHIST OFFERINGS 365 DAYS

STYLE ELLE DECO
Vivre au Maroc
VILLAS ET PALAIS DE ROME

Jackson Pollock

modern sculpture can stand beside classical bronzes. Plus the home shows the work of a couple who, in collecting terms, are always one step ahead of the game: Thong describes a bronze-coated plaster sculpture he bought from an artist who, at the time, was not well known, but whose work now is highly collectable. "I just bought it because we loved it," he recalls, "but it's always satisfying to be one of the first to spot an artist's potential."

The library, Anne's domain, bears witness to her love of textiles and finishes, from the green of the intricate hand-blocked gauffrage fabric – built up using pattern upon pattern – that covers the 1940s'-style chairs, to the Dior-pleated silk shades at the desk. Attention to detail characterises her work: it's no surprise that a cushion on the sofa in the living room isn't covered in any old velvet but, rather, in a hand-made, hand-dyed design that has a unique sheen and texture. In the bathroom, too, it's Anne's capacity to soften a look that stops the space from looking too understated. The simple, classic design, with its contemporary take on the Corinthian column and lineal geometric patterning continuing from the floor into the shower shows off the room's lofty proportions and marble finish to its best advantage. So far, so good, but the unpredictable element that really grabs the attention comes courtesy of a magnificent Baroque-style gilded console table, home to glass jars of shells and fossilised corals. Because the table isn't what one would expect to find in a bathroom, it's what makes the space truly special, and it works because its elaborate design provides a terrific contrast to the pared-down look of the rest of the room.

Decoration Empire's founders would be the first to say that where their own home differs from the projects they undertake for clients is that they've had some 15 years to let their designs take shape and come to fruition – projects for clients tend to be completed far more quickly! "In some ways, this home reflects the history and growth of our business because we bought it 15 years ago, when Decoration Empire was relatively new," says Thong Lei. "But as we've seen our company grow and our experience and horizons broaden, we've added to the apartment, bit by bit. Every home should grow organically, and the best ones do so automatically: their owners' love for beautiful things means that new treasures are regularly added, design re-thinks are undertaken on a regular basis. That's how the loveliest homes should evolve, and how they can best reflect the lives of those they shelter." |

水
江

永
平安桃巷坊
金谷女兒國松都美五歲書
樂
國松都美五歲書

XII

Another look at the key pieces
featured within each of the apartments,
details of the craftsmen, artists
and suppliers behind them,
plus the provenance of many
of the most covetable antiques...

I

Clockwise from top left: a crystal lamp with Honan silk Dior pleated lampshade; Toscana armchair upholstered in mauve velvet from Manuel Canovas, with African treasures on the side table; a Neolithic jade bi-disc from China, on a bronze standard. The true significance of these discs is unknown, but they must have had an important ritual function as part of the burial; a pair of 18th-century Chinese porcelain lions; an artwork made from a small bowl set within the pages of an old red book; male and female bronze statues from the owners' private collection; a bronze of Julius Caesar sits aside a 19th-century bronze Buddha from Thailand; detail of a Murano glass dish.

I

Clockwise from top left: detail of an 18th-century Chinese table with hand-carved and gilded embellishments; a dining chair from the Empire Home Collection upholstered in Hopsack fabric in Sungold, with oak table made to measure by Decoration Empire; the kitchen cabinets are of Havana oak, with honed black stone surfaces and a chrome tap; the china was bought in various Japanese markets; Louis Durot's 1970s' polyurethane chair *La Spiral*; the dressing room was designed by Decoration Empire with cupboard doors in two-colour Zebrano wood, plus brass door handles by Chloe Alberry; the bath is topped with Italian *Madre Perla* marble from Antollini; detail of a Dornbracht single-lever Lulu tap in chrome.

III

Clockwise from top left: a French 19th-century wooden urn with hand-carved garland has been converted into a lamp; bespoke panelled doors made by Decoration Empire have been handpainted to resemble oak; this 18th-century Gustavian armchair from Sweden has been upholstered in linen; a wardrobe custom-made using Italian frescoed doors set into hand-carved frames; a pair of 18th-century Swedish Gustavian footstools upholstered in fabric from Georges Le Manach's *Carreaux Rouge* collection; black-and-white striped wallpaper in the hall by Ralph Lauren complements a simple wall-mounted light fitting; bespoke flooring by Decoration Empire is made from hand-cut and polished Belgian bluestone; Porta Romana's bronzed Oliver wall lights, topped with semi-circular silk shades.

III

Clockwise from top left: the library features hand-carved pine panelled walls, designed, installed and painted using a special finish by Decoration Empire; 18th-century French oak parquet flooring in *point d'hongrois* by Decoration Empire; a single-arm wall light in antiqued brass with pleated red silk shade in the library; rattan dining chair upholstered in Ziro by Malabar; these cabinet doors have been coated with six layers of gesso and paint following a traditional method for making picture mouldings; open storage in the bathroom has been created from an antique arch rescued from an old Italian theatre and restored by one of Decoration Empire's craftsmen; this brass-finished wall light from France has been fitted with a halogen bulb; a bespoke headboard in distressed oak, designed by Decoration Empire, has been upholstered using Zoffany's Broccatello fabric.

V

Clockwise from top left: a porcelain Dutch tulip vase sits on an oak-topped coffee table; a 17th-century portrait of a lady by Jacob Willemszoon Delff; detail of a 17th-century cabinet inlaid with ebony and ivory; a Chinese terracotta horse from the Tang dynasty; a marquise bergère chair, painted in beige and grey and upholstered in fabric by Clarence House; American wooden shutter with 9cm wide louvre on a movable system from Palm Shutters; cabinets designed by Decoration Empire and painted in matte lime paint, plus bone doorknobs by Chloe Alberry; nickel light fittings with bespoke silk shades hang on panelled walls created and installed by Decoration Empire.

V

Clockwise from top left: 60cm x 60cm hand-cut and polished Belgian bluestone floor tiles edged with oak inlay; the wall in the shower features Italian *fior di bosco* marble with a band of white *statuario* marble, both from Antollini; bespoke carpet in finest wool designed by Decoration Empire; a dark green glass 1950's lamp with Honan silk Dior shade sits on a black-lacquered side table from the Empire Home Collection; an old fish tooth from Denmark contrasts with an antique Chinese table; in the guest bathroom, a Geminus tub from The Albion Bath Company has been painted in *gris chaud* matte paint from the French brand Ressource; these multicoloured 10cm x 10cm ceramic floor tiles are from the French company Winckelmans.

VII

Clockwise from top left: French chair from the Milling Road collection by Prelle; detail of a 1969 painting by Jean Pierre Pincemin, plus a 19th-century empire column chandelier that has been converted into a lamp and two Khmer civilisation torsos, dating from the 15th to 16th centuries; a bronzed sconce featuring a Blackamoor head and mounted on antiqued mirrors; a group of terracotta musicians from the Chinese Han Dynasty; on the antique carpet stands an English 18th-century hand-carved and gilded side table with a Chinese lacquered top; *Feuervogel* sculpture by Hungarian artist Laszlo Szabo; painting by the Dutch artist Miquel Ybanez viewed through an old mirror glass; marble objects of the third and early second millennium BC, originating from in the Bactria-Margiana area in the northern region of Afghanistan and the Persian province of Margush.

VII

Clockwise from top left: detail of an antique footstool upholstered in Giraffe velvet fabric by Edmond Petit; in the dressing room, detail of bespoke, hand crafted doorhandles in distressed brass, designed by Decoration Empire; this antique table lamp of multicoloured Murano glass dates from around 1950; the dressing-room doors feature antique mirrored glass set into distressed brass frames; black sidetable in solid oak from the Empire Home Collection; in the bedroom, panelled walls are upholstered in Paso horsehair fabric from John Boyd Textiles; this white-onyx-topped washstand from the Empire Home Collection features two mirrored drawers plus crystal doorknobs by Chloe Alberry, the tap is Madison Flair by Dornbracht; the bathroom floor is of Italian *Fior di Bosco* marble from Antollini with an inlay of platinum mosaic stone from the Murano Smalto collection by Sicis.

IX

Clockwise from top left: dining chairs from the Empire Home Collection upholstered in an Italian linen-cotton mix, around a Spanish dining table; a footstool, designed by Anne Noordam for Decoration Empire, features gilded lion's claw feet and is upholstered in Boucle Zebra by Clarence House; this Gouda wing chair is from the Empire Home Collection, with a loose cover of finest wool, on the desk sits an ostrich egg; detail of the bedside cabinet in walnut; this bespoke headboard is upholstered in a Manual Canovas fabric, with integral single-arm wall light plus linen shade; a 19th-century stone statue from Cambodia; a 19th-century cast-iron inkpot featuring Napoleon; the bronze and marble table dates from 1940, on top of it sits a contemporary sculpture in the hall by an artist from the South of France.

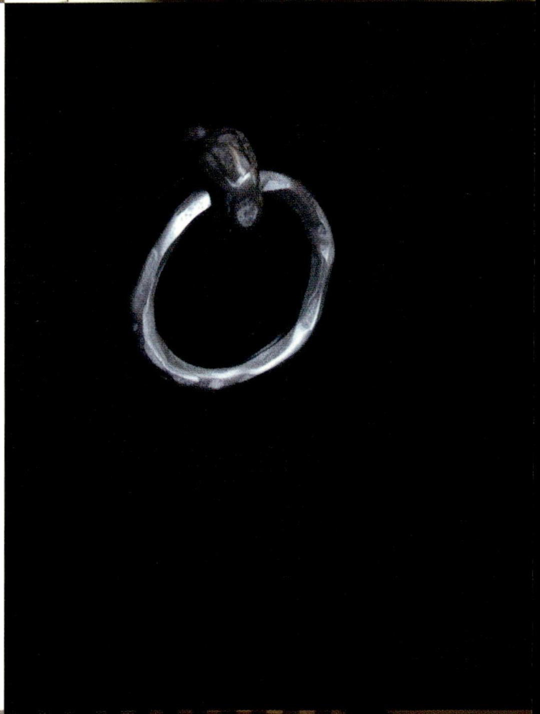

IX

Clockwise from top left: a collection of grey terracotta Chinese jars from the Han Dynasty period; a detail of a mixed media painting by the Dutch artist Wil Jansen; her study, with a floor of dark-stained oak by Decoration Empire and a painted metal bookcase in bright yellow and chrome by USM; the brushed aluminium and leather desk chair is from Hadi Teherani's Silver range; view of Frank Gehry's Wiggle Chair, designed in 1972 and made from 60 layers of corrugated cardboard; acrylic on canvas work by the Spanish artist Eduardo Barco; large black and white vase from the Empire Home Collection; double-arm wall light in nickel – hidden within the silk shade is a special LED light to facilitate easier reading.

XI

Clockwise from top left: a torso from the Khmer dynasty stands in front of a painting by François Millon; a hand-made cushion in dark turquoise silk softens the Gouda wing chair from the Empire Home Collection; the sofa is upholstered in brown *velour de soie* by Tassinari, on the table in front sits a 19th-century wooden sculpture of the Leaning Tower of Pisa and a marble stone from the Bactria-Margiana area, early second millennium BC; an early 19th-century footstool by the French furniture maker Jacob Desmalter is upholstered in green mohair Villandry gauffrage velvet; detail of a painting by the Dutch artist Bram Bogart; a pair of French bronze gilded candleholders dating from the Napoleon III era; painting by French artist Marie-Cécile Aptel and a cement sculpture by Alexandra de Lazareff; a pair of 19th-century Chinese vases plus a collar made of shells from the Oceania region.

XI

Clockwise from top left: a 20th-century travel journal in gouache by the French artist Bourguin; filing cabinet in chrome and bright red metal by USM; table lamp, designed by Wilhelm Wagenfeld in 1924 for Technolumen; painting by the Dutch artist Fred Belmer; canopy made of dark green silk lined with pink, with gold tiebacks and a reading light in antiqued brass; Koningsbourg wallpaper by Zuber Cie plus antique *intaglio* seals made of alabaster and displayed in a hand-made wooden frame; headboard in Maestro silk velour by Lelièvre, with Hautefort gauffrage in a hand-carved and gilded frame with a Regency tieback beyond; Italian fossil limestone marble in Lucido from Antollini, plus a modern bronze by the Dutch artist Miguel Ybanez.

Credits

©2009
Uitgeverij Terra Lannoo BV
Postbus 614, 6800 AP Arnhem
info@terralannoo.nl www.terralannoo.nl
Uitgeverij Terra is part of the Lannoo-group, Belgium

Text
Anna Lambert,
United-Kingdom

Photography
Thong Lei
Gouda

Styling
Anne Noordam
Gouda

Styling assistance
Jolanda Eegdeman and Madelon Overeijnder

Art direction & design
mennolandstra.com
Easterein

Lithography
Veronica Litho
Hilversum

Printed and bound
Printer Trento
Trento (Italy)

With thanks to:
Co van der Horst, Chanel , Sissy-Boy, Waterproof Bathrooms, Marc Smithuis Keukenarchitectuur, Antique Express, Gimko International, Gert Snel, ISPC, de Tafeldame, de Spiegeltent Gouda. The Decoration Empire team, whose members are always jointly responsible for project execution.

ISBN 978-90-8989-012-2 (Dutch edition) | ISBN 978-90-8989-038-2 (English edition)
NUR 454